29.93

# EXPLORING MARS

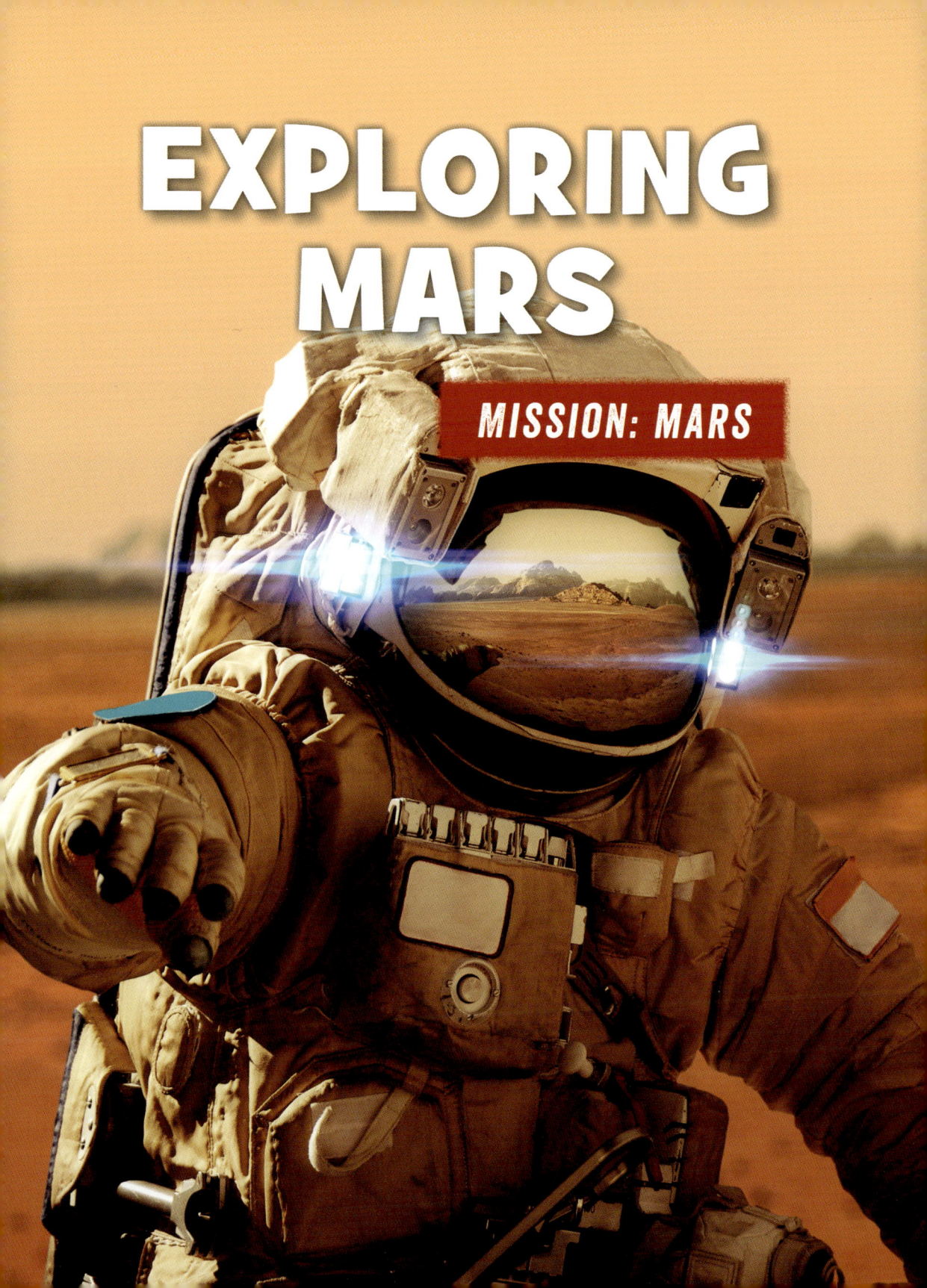

## MISSION: MARS

# Cherry Lake Press

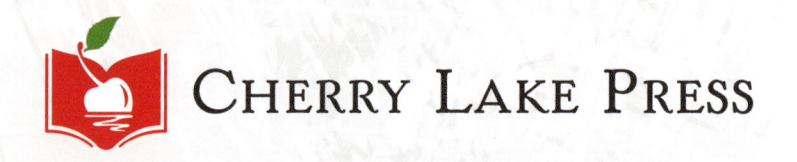

Published in the United States of America by Cherry Lake Publishing
Ann Arbor, Michigan
www.cherrylakepublishing.com

Reading Adviser: Beth Walker Gambro, MS, Ed., Reading Consultant, Yorkville, IL
Book Designer: Book Buddy Media
Photo Credits: Cover: ©Sergey Nivens / Shutterstock; page 1: ©Alones / Shutterstock; page 5: ©Think_About_Life / Shutterstock; page 7: ©by Marc Guitard / Getty Images; page 9: ©Miriam Espacio / pexels.com; page 10: ©Dotted Yeti / Shutterstock; page 15: ©NASA/JPL/Corby Waste / Wikimedia; page 17: ©ESA/DLR/FU-Berlin / nasa.gov; page 19: ©NASA/JPL-Caltech / nasa.gov; page 21: ©3DSculptor / Getty Images; page 23: ©NASA/JPL-Caltech/MSSS / Wikimedia; page 25: ©u3d / Shutterstock; page 27: ©Cavan Images / Getty Images; page 29: ©MARK GARLICK/SCIENCE PHOTO LIBRARY / Getty Images; page 30: ©nasa.gov;

Copyright ©2022 by Cherry Lake Publishing Group
All rights reserved. No part of this book may be reproduced or utilized in any form or by any means without written permission from the publisher.

**Cherry Lake Press** is an imprint of Cherry Lake Publishing Group.

Library of Congress Cataloging-in-Publication Data has been filed and is available at catalog.loc.gov

Cherry Lake Publishing would like to acknowledge the work of the Partnership for 21st Century Learning, a Network of Battelle for Kids.
Please visit *http://www.battelleforkids.org/networks/p21* for more information.

Printed in the United States of America
Corporate Graphics

## ABOUT THE AUTHOR

Mari Bolte is a children's book author and editor. Streaming sci-fi on TV is more her speed but tracking our planet's progress across the sky is still exciting! She lives in Minnesota with her husband, daughter, and a house full of (non-Martian) pets.

# TABLE OF CONTENTS

**CHAPTER 1**
## Destination: Mars .................................. 4

**CHAPTER 2**
## Traveling Through Space ................... 8

**CHAPTER 3**
## X Marks the Spot ............................. 14

**CHAPTER 4**
## Exploring the Planet ........................ 18

**CHAPTER 5**
## The Future ........................................24

ACTIVITY ................................................................ 30
FIND OUT MORE ...................................................... 31
GLOSSARY .............................................................. 32
INDEX...................................................................... 32

# CHAPTER 1

# Destination: Mars

Besides Earth, Mars is the most-studied planet in our solar system. Earth and Mars share many similarities. Both planets are terrestrial, which means they are made of rocks or metal. It takes them similar amounts of time to rotate around the Sun. And, like Earth, Mars has shown signs of having water on its surface.

The challenge of visiting, or even colonizing, Mars is one that astronauts look forward to facing head-on. Scientists have sent robotic explorers called rovers there to learn as much as they can. But what is there to see on Mars that we can't see here on Earth? Turns out—a lot!

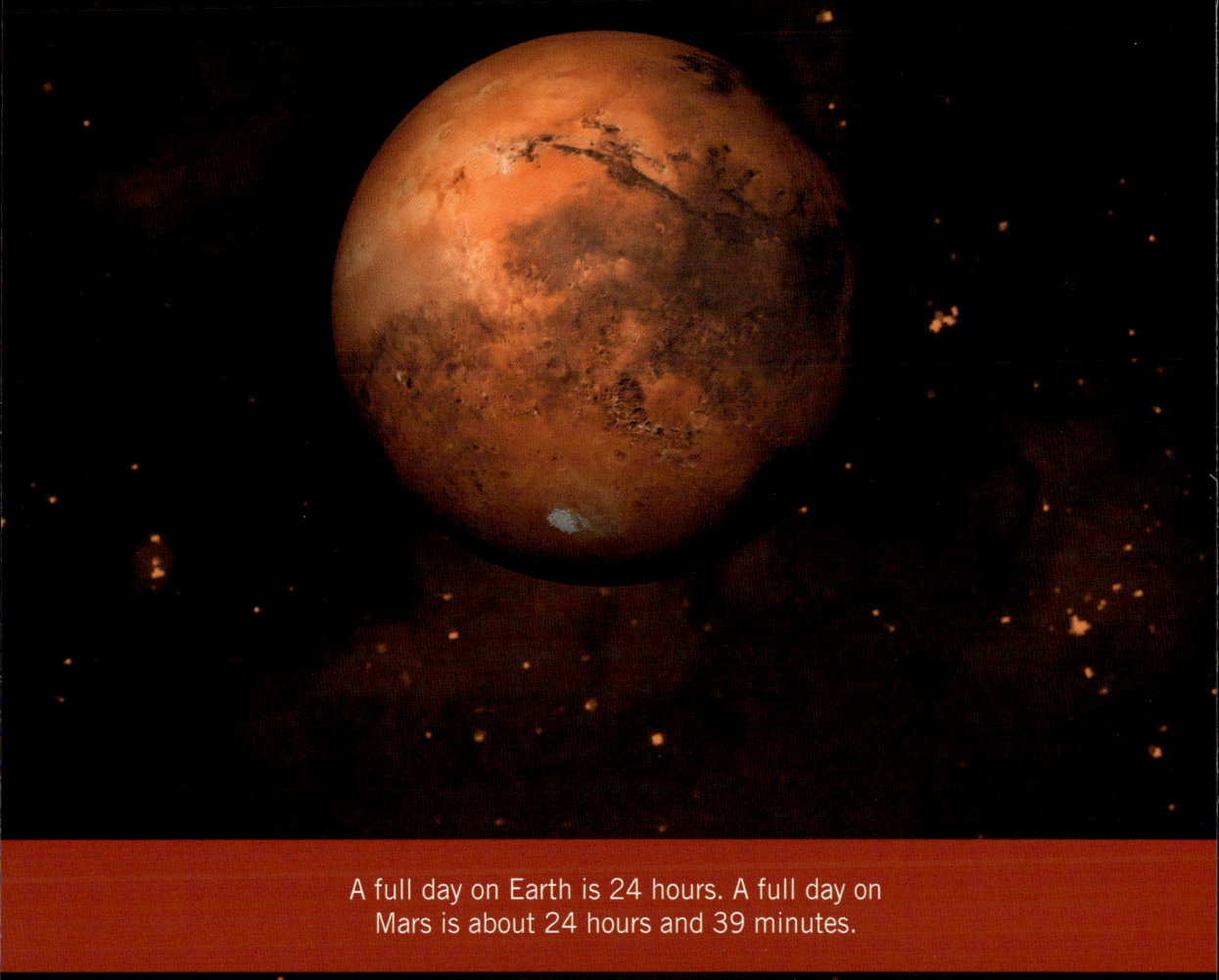

A full day on Earth is 24 hours. A full day on Mars is about 24 hours and 39 minutes.

## Seeing Is Believing

The first telescope was built in the early 1600s. In 1610, **astronomer** Galileo Galilei was the first person to see Mars through a telescope. Today, even small, basic telescopes will let you glimpse the Red Planet. But telescopes are so 20th century! The Mission to Mars AR mobile app, released by the Smithsonian Channel in 2021, allows people to experience Mars through augmented reality. Driving rovers, launching rockets, and completing missions are only a few interactive ways to visit space without actually going there. The Portal to Mars feature lets the user walk across a red desert while the entry back to their living room shrinks in the distance.

When scientists find interesting geographical features on Mars, they name them after an equally interesting place on Earth. The Martian Monument Valley is named after Monument Valley in Arizona and Utah. Both are made up of **mesas** and **buttes**. Their hills are layered rock that have been **eroded** away. The mesa tops are flat, hard sandstone. Photos of the Mars surface have been compared to photos of the American southwest taken in the late 1800s.

The tallest planetary mountain in the solar system, Olympus Mons, stands more than two and a half times taller than Earth's highest point, which is Mount Everest. Olympus Mons stretches 15.5 miles (25 kilometers) above the planet's surface. It is 374 miles (602 km) wide at the bottom. It is one of many

Lava from shield volcanoes pours out of vents or cracks and is pushed outward. Overflowing layers of lava create the shield volcano's gently sloping sides.

**shield volcanoes** on the planet. Wider than they are tall, these volcanoes look like a warrior's shield.

Mariner Valley has been called the Grand Canyon of Mars. It dwarfs Earth's canyon, though. The Grand Canyon on Earth is about 500 miles (800 km) long. Mariner Valley is 2,500 miles (4,000 km), which is almost as long as the entire United States!

Mars's Chaos Terrain is like nothing on Earth. Scientists describe its fractured appearance as a flat piece of chocolate that has been frozen and then smashed. It is believed to have been formed when ice froze in **craters**. Then the ice melted and the water flowed away, causing the ground to erode or collapse.

## CHAPTER 2

# Traveling Through Space

Leaving Earth and flying to Mars may sound simple. But there is a lot more to it than piloting a spacecraft! Both planets **orbit** the Sun, but they orbit at different paces. It takes Earth 365 days to make a loop around the Sun. It takes Mars 687 days to do the same thing. Sometimes the planets are far apart, on opposite sides of the Sun. Sometimes their orbits bring them close together.

# Mars Orbit

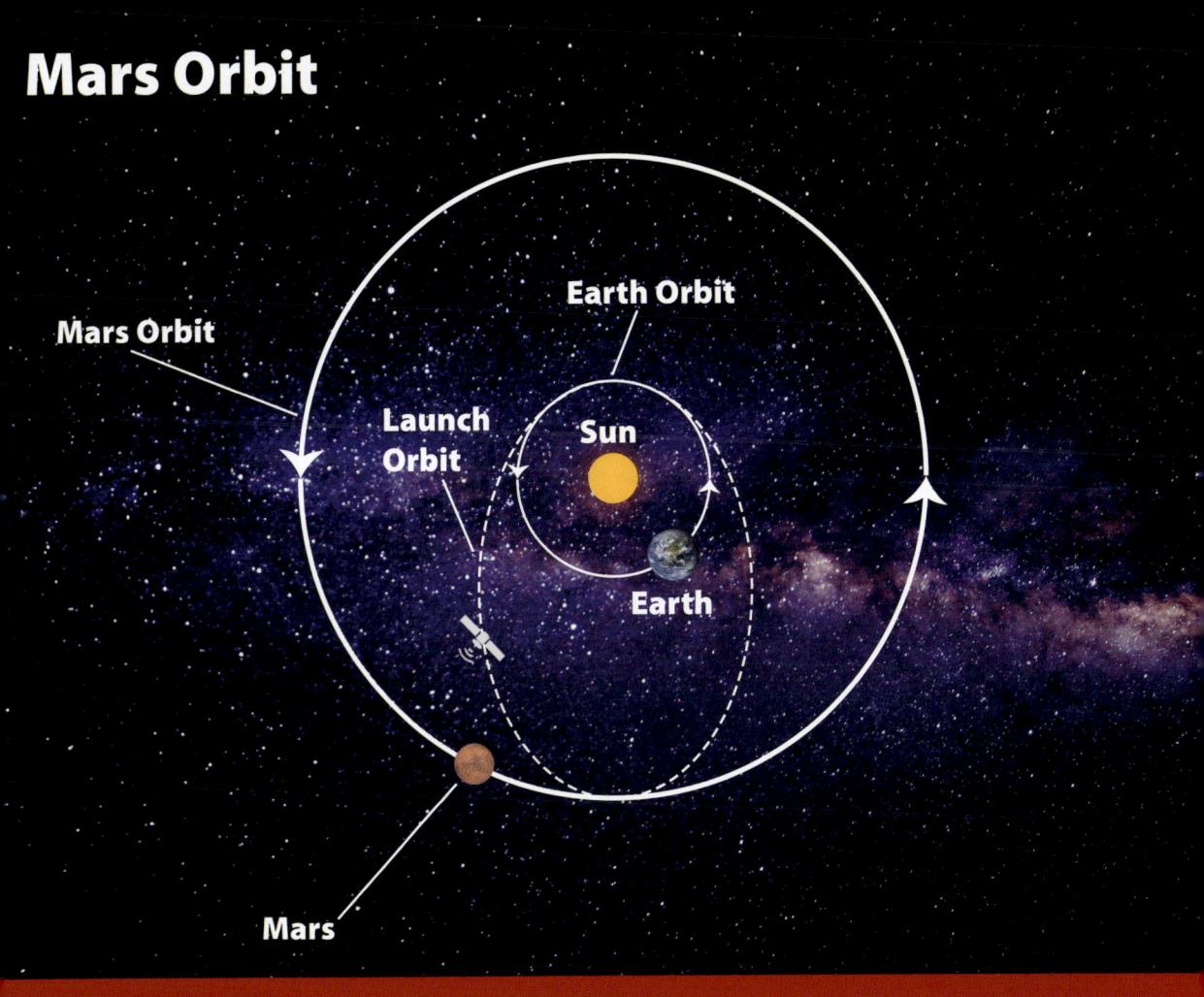

*An orbiter traveling on an elliptical orbit might follow that path two or three times before lining up with Mars.*

The planets' orbits are two different shapes. Earth's is fairly circular. Mars's orbit is more egg-shaped. This means that Earth and Mars are anywhere from 33.9 to 250 million miles (54.6 to 401 million km) apart, with 140 million miles (225 km) the average distance.

When Mars, Earth, and the Sun line up, it is called opposition. Mars shines brightest during opposition.

With such a big difference between the planets' distances, leaving at the right time can be a challenge. When you think of how far apart the two planets can be, it makes sense that astronauts would want to launch at the most ideal time. But timing the landing on Mars isn't easy. The planets line up every 25 to 26 months, and it will take at least 7 months of travel on a spacecraft to reach Mars.

## Surviving the Journey

*Seven months is a long time to be in space. Then, add on the mission time and the flight back. Some have compared it to half a dozen people riding in an RV together for 3 years. How will people cope with the longest car ride in the universe? Some of the techniques being used on the International Space Station (ISS), like fresh fruit deliveries, window views, and conference calls back home, will be much harder, or even impossible, on the trip to Mars. Virtual reality (VR) is one idea that could help astronauts cope. Pairing VR with exercise, musical instruments, or even just something different to look at could help keep astronauts sane.*

Because of the planets' orbits, the 400 million (644 million km) mile trip won't be a straight flight. The spacecraft would need to be launched in an **elliptical** arc to intercept Mars later in the planet's rotation. Calculating the path to the landing site on Mars is complex, and data can change. Deciding whether to take the fastest route or the route that uses the least amount of fuel is another thing to consider.

## Skills on Mars

*Getting around on Mars will not be as easy as it is on Earth. Think about all the things that make it easy to go places here—roads, lots of cars, and breathable air. None of those exist on Mars yet! Space taxi drivers and delivery drivers will be needed to help move tourists and colonists around. The National Aeronautics and Space Administration (NASA) is already working with private companies to use drones to shuttle people and packages from Earth to the ISS. Even if the space taxis had automated drivers, someone would need to service and maintain them. Space mechanics could be an essential job.*

When you look at the sky, it looks like there is nothing between here and Mars. But there is more than just empty space between the two planets!

Earth has one moon, but Mars has two. They are named Phobos and Deimos. They are much smaller than Earth's Moon. Phobos is 14 miles (22.5 km) across, and Deimos is around 8 miles (12.9 km) wide. To compare, Earth's Moon is nearly 2,160 miles (3,476 km) across. Made of carbon-based rock and ice, they are similar to **asteroids**. They are lumpy and cratered, which makes them look even more asteroid-like.

# Orbiting Mars

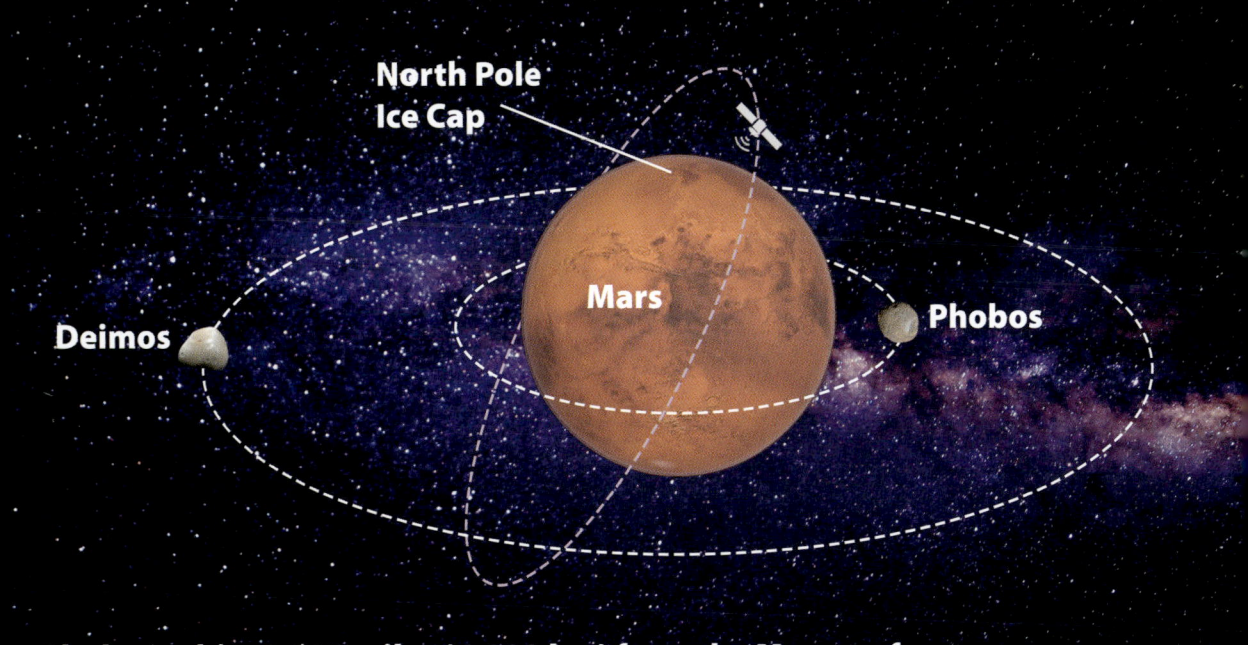

North Pole Ice Cap

Deimos

Mars

Phobos

Phobos orbits 3,700 miles (6,000 km) from the Mars surface.
Deimos orbits 14,580 miles (23,460 km) from the Mars surface.

Phobos orbits around Mars three times a day. It takes Deimos 30 hours to do the same thing.

Phobos could be a good base location for astronauts studying Mars. The moons could also serve as a landing site. People could stop on Phobos or Deimos and then be ferried to the surface of Mars.

CHAPTER 3

# X Marks the Spot

There have been 12 attempts to actually land spacecraft on Mars. Ten were successful. Landing sites were chosen with the goals of land, work, and discover. Successful landing sites were places without high elevation. The planet's atmosphere is thinner the higher you go, and that can affect a landing. Some unsuccessful landings have been due to the atmosphere affecting when parachutes and airbags **deploy**. Also, steep slopes, dust storms, or big rocks could damage both the landing equipment and the rovers. The first spacecraft to successfully land on Mars, the former Soviet Union's *Mars 3*, touched down on December 2, 1971. But 20 seconds later, it went dead. A dust storm on the planet's surface is believed to be the cause.

*Phoenix* was launched on August 4, 2007. It landed on Mars on May 25, 2008.

Working near the planet's equator will allow for more consistent weather patterns. It also provides more sunlight for solar equipment. In 2007, the NASA **probe** *Phoenix* was sent on a three-month mission. Its goal was to figure out if the planet had ever had life, and if it could ever support new life. Thanks to solar energy, it was able to keep working for five more months. Eventually, reduced sunlight caused it to shut down.

## Risky Business

*Exploring a new planet has risks. The main risk right now is losing spacecraft. But once people are involved, the risks become deadly. One mistake or accident can lead to injury or death. And human bodies are soft. Space suits will give them the protection they need. Greater mobility will let people move around more easily for exploration or outside repairs. They will maintain a comfortable temperature even in extreme* **climates**. *In an emergency, the suits can be used as safety shelters for up to 6 days. A portable life support system recycles air and removes odors and moisture from the suit.*

Finally, some landing sites offer different things to discover. One location might look more hospitable for future colonies. Another might have interesting rock formations nearby. Between 2006 and 2011, scientists got together to discuss missions on Mars. They made a list of 100 potential places to land. Then they narrowed the list down to four. The four locations all had clear evidence of being able to support life, either in the past or present. They had geological areas that would be easy to study, such as exposed layers of rock or dried-up lakes. And they would be easy and safe places to land.

Scientists believe the Jezero Crater was flooded long ago. Rivers carried water over the crater's wall, filling it with water and making a lake. Clay minerals that are carried by water were found in the crater.

In 2021, the *Perseverance* rover landed in Jezero Crater. A rover is a small vehicle designed to drive across the surface of Mars. NASA had spent 5 years poring over 60 potential sites before choosing Jezero. They believed Jezero could have supported life 3.5 billion years ago. When the planet's atmosphere grew too thin, its water disappeared. *Perseverance* is now looking for signs of past life in the crater's rocks and soil.

# CHAPTER 4

# Exploring the Planet

We have sent more spacecraft to Mars than any other planet. The first **flyby** took place in 1965, when the probe *Mariner 4* took the first photos of Mars. Since then, several organizations and countries have tried to make it there. Until 2021, NASA was the only organization that has actually operated spacecraft on the planet's surface. That changed on May 14, 2021, when China's *Tianwen-1* orbiter delivered their rover, *Zhurong*, onto the planet's surface.

*Mariner 4* was built in NASA's Jet Propulsion Laboratory (JPL). JPL operates out of the California Institute of Technology, or Caltech.

The only risks so far have been the loss of spacecraft. Sending humans to Mars would come with real danger. The long flight would be hard on human bodies and minds. But there are other risks too. Earth's magnetic field protects us from space **radiation**. Radiation is all around us. It is a form of energy that can have negative effects on our bodies. But space radiation is even more dangerous. Long exposure to space radiation can cause damage to the central nervous system. This could include dizziness, body coordination problems, or damage to the memory. It also can cause cancer.

> ## Deep-Sea Dives
>
> *In 1872, the ship HMS* Challenger *set out on a 1,000-day scientific mission. It carried great lengths of rope to record the ocean's depths. It also took samples of silt and water at those depths and used special nets to collect animals from the bottom of the sea. Many organisms had never been seen before. The ship visited every continent except Antarctica. It discovered the deepest part of the ocean, a place we know today as Challenger Deep. The space shuttle* Challenger *was named for the ship and its explorers.*

Once on the planet, there would be more danger. Mars's gravity is three-eighths the amount of Earth's. Being lighter sounds fun, but it is surprisingly hard on the human body. Not needing to use all of your strength can cause your muscles to weaken. Astronauts can lose up to 20 percent of their muscle mass in just 5 to 11 days in space.

Colonizing Mars would mean living in a closed environment. Things like temperature, air quality, and light would need to be constantly monitored. Food would be limited. What if someone got sick? Their germs would be easily passed from one person to the next.

*Challenger* was the second shuttle to reach outer space, in 1983. It made nine successful launches.

## By the Numbers: Close Approach and Opposition

*Close Approach is the point when the path of Earth and Mars's orbits places the planets near each other. In 2003, Mars and Earth were less than 35 million miles (56 million km) apart. This was the nearest to each other that they had been for 60,000 years. They will not be that close again until 2287. October 6, 2020, was the last Close Approach. On that date, the planets were about 39 million miles (62 million km) apart. Opposition is reached when Mars, Earth, and the Sun line up. It happens about every 2 years and 2 months, or every 780 Earth days. From Earth, Mars rises in the east sky at sunset. It is at its brightest at midnight. Then it sets in the west at sunrise.*

Finally, the planet's surface would pose more threats. The dust storms on Mars are so huge that they can be seen through telescopes. Every 5.5 Earth years, the storms get so big that they cover the planet. While the dust does not blow hard enough to do damage, the dust particles themselves are a problem. The small particles are full of static electricity and like to stick together. Machines do not do well while covered with dust. And dust covering solar panels can keep them from collecting energy.

The rover *Curiosity* made its first drive across Mars on August 22, 2012. This photo was taken by *Curiosity*.

Mars is covered in boulders, mountains, canyons, and craters. The rough terrain can damage rover wheels. It can also be dangerous to anyone hiking across the surface. Not only could falling on sharp rocks hurt, but the rocks also could rip or damage a space suit.

CHAPTER 5

# The Future

The first humans on Mars will have a lot of responsibility. The first thing they will need to do is set up a base. Next, they will need to establish a 60-mile (100 km) area around the base. This exploration zone will be a research site. The people would also find things that could be harvested for use.

People on Earth have used mud, rocks, and clay to build shelters. The dust on Mars could be used in a similar way, to make Martian concrete. It could also be compressed to make bricks. It could even be turned into a material for 3D printing.

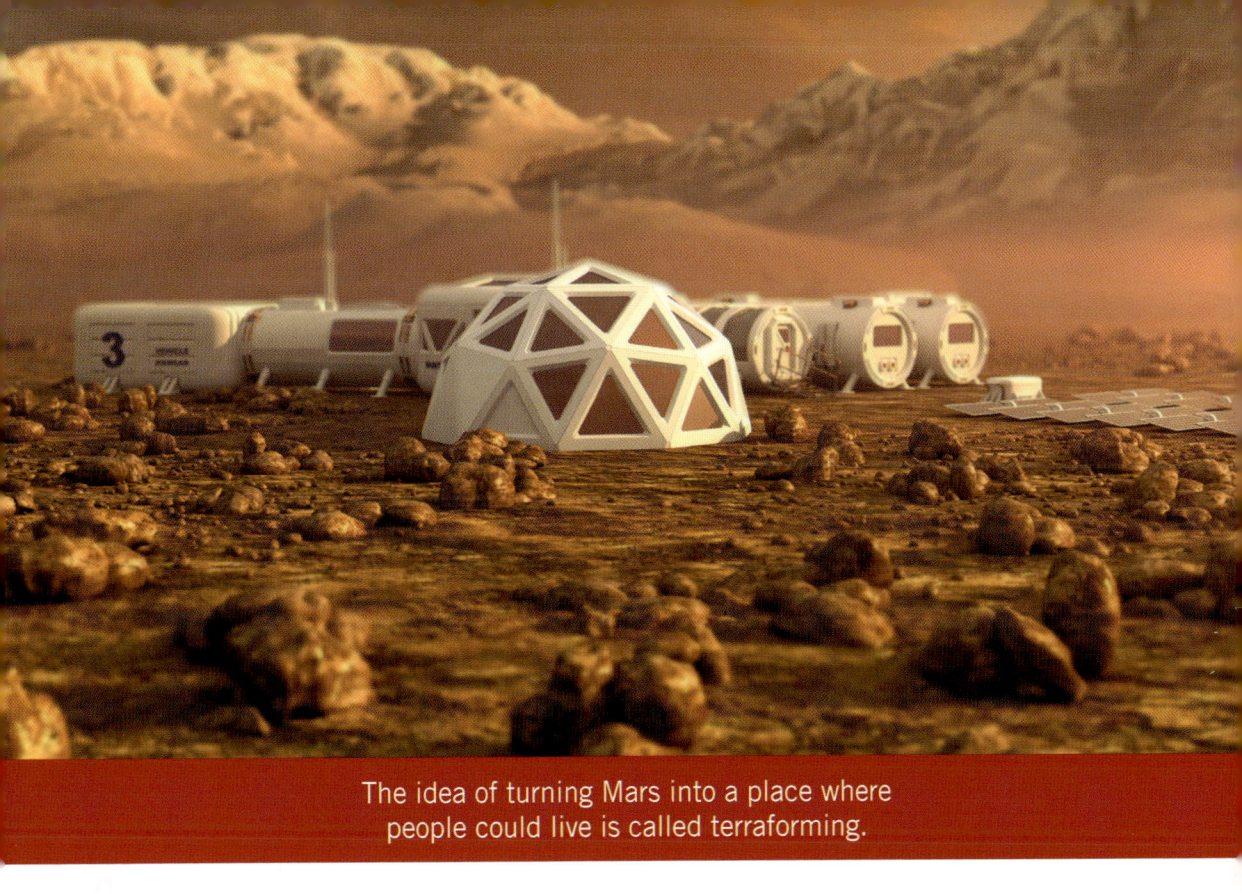

The idea of turning Mars into a place where people could live is called terraforming.

Astronauts, doctors, and **geologists** would need to work together to **colonize** Mars. But they would need farmers too. Mars has some of the right nutrients to help plants grow. And plants grown on Mars would need less water. This is because of Mars's reduced gravity. The soil could hold more water, and the water would not drain away as quickly as on Earth.

Supply missions could be launched every year. They would arrive at the planet every 2 to 3 years. Building supplies to expand the base would be part of these missions. Where would those bases be built? Living underground makes the most sense, protecting explorers from dust storms, extreme temperatures, and radiation.

Robots could dig caves for underground habitats, or places to live, before people arrive. Volcanic caverns that were formed by lava are another possible site. If the caves connect, they could provide already-made paths between bases. They might also hold trapped ice.

## Living Responsibly

*When deciding on a landing site, NASA looks for places **microbial** life may still be present. Then they avoid those places. If there is anything alive on the planet, it is very important that we do not damage it. There would be a limited amount of resources. Water, oxygen, and even waste would need to be recycled or reused. Garbage is another consideration. All the old rovers on Mars are still there. They might eventually be picked up by humans and put in a museum. Or they might stay buried for people to discover millions of years in the future.*

Volcanic caves are usually not far from the surface. They can stretch on for miles.

[ Mission: Mars ]

Billions of years ago, Mars was in motion. Volcanoes erupted. Lakes flowed through valleys. The planet's surface, which today reaches temperatures as low as −220 degrees Fahrenheit (−140 degrees Celsius), was warm enough to support life. It sounds strange. But Earth supported early life 3.8 billion years ago. It is possible Mars had life too. Some places on Earth have even been used to study Mars. Antarctica, South America, and Hawaii all have conditions similar to the Red Planet. If it is proven that Mars had life, even microbial life, it would mean it could host life in the future too.

## Take a Trip

*Still want to hop on a spacecraft headed for Mars? Get ready for the longest road trip ever! You'll have to bring every item of clothing, piece of food, or drink of water you'll need during your 7-to-9-month journey. What you bring to entertain yourself needs to last the entire time. There's no Wi-Fi, and space radiation could fry your phone. There won't be any stops at the gas station for snacks. If you forget something, there will be no turning around! Did you bring enough toothpaste?*

Olympus Mons is in the Tharsis Montes region of Mars. Volcanoes in this area are 10 to 100 times larger than volcanoes on Earth.

# Activity: Pack Light

What would you take to Mars? Today, astronauts on the way to the ISS are allowed to bring 3.3 pounds (1.5 kilograms) of personal equipment along. But how much is that, really? Think about what you would want to have on Mars. What is important to you? What would you really miss?

## WHAT YOU'LL NEED:

- favorite items
- scale
- pencil and paper

1. Weigh each item on the scale. Record its weight.
2. Continue until each item is recorded.
3. Figure out how many items you could bring along. Remember, the total weight for everything needs to be under 3.3 pounds (1.5 kg)!

# Find Out More

## BOOKS

**Huddleston, Emma.** *Explore the Planets.* Minneapolis, MN: ABDO Publishing, a division of ABDO, 2021.

**Lomberg, Michelle.** *Mars.* New York, NY: AV2 by Weigl, 2019.

**Stuart, Colin.** *How to Live in Space: Everything You Need to Know for the Not-So-Distant Future.* Washington, DC: Smithsonian Books, 2018.

## WEBSITES

### Britannica: 10 Important Dates in Mars History
*https://www.britannica.com/list/10-important-dates-in-mars-history*
Discover 400 years of Mars research, from the very first glimpses to the discovery of water.

### Discovery Education: Explore Space
*https://www.discoveryeducation.com/learn/explorespace*
Take a virtual trip to Mars and watch as *Perseverance* touches down on the Red Planet.

### NASA's Mars Exploration Program
*https://mars.nasa.gov*
Experience up-to-the-minute updates about the Red Planet, from weather to recent discoveries.

# GLOSSARY

**asteroids** (AS-tuh-roydz) small rocky bodies in space

**astronomers** (uh-STRON-uh-muhrz) people who study space

**buttes** (BYOOTS) hills with steep sides and flat tops; buttes are smaller than mesas

**climates** (KLY-muht) the weather conditions in an area over a long period of time

**colonize** (KOL-uh-nyz) to send a group of settlers to a new place

**craters** (CRAY-tuhrs) large bowl-shaped depressions in the ground

**deploy** (duh-PLOY) to move something into position

**elliptical** (uh-LIP-tuh-kuhl) an oval, egg-like shape

**eroded** (uh-RODE-uhd) worn away gradually by natural means, such as water or ice

**flyby** (FLY-by) sending a spacecraft within close range of a planet

**geologists** (gee-AH-luh-jists) scientists who study the makeup of Earth and other planets

**mesas** (MAY-suhz) flat-topped hills with steep sides

**microbial** (my-KROW-bee-uhl) the presence of very small living things, called microbes; microbes include bacteria, algae, and amoebas

**orbit** (OR-bit) the curved path of an object in space around a star, planet, or moon

**probe** (PROHB) an unpiloted spacecraft that travels through space to collect information

**radiation** (ray-dee-AY-shuhn) a form of energy that travels through space

**shield volcanoes** (SHE-uhld vol-KAY-nohz) volcanoes that are wide across and have shallow, sloping sides

# INDEX

geological features on Earth
    Grand Canyon, 7
    Monument Valley, 6
    Mount Everest, 6

geological features on Mars
    Chaos Terrain, 7
    Jezero Crater, 17
    Mariner Valley, 7
    Martian Monument Valley, 6
    Olympus Mons, 6, 29

HMS *Challenger*, 20

International Space Station (ISS), 11, 12, 30

*Mission to Mars AR* app, 6
moons, 13

orbits, 8, 9, 11, 13, 22

rovers, 4, 6, 14, 26

spacecrafts
    *Mariner 4*, 18, 19
    *Mars-3*, 32
    *Perseverance*, 17, 31, 32
    *Phoenix*, 15, 32

virtual reality (VR), 11